ENHANCED WEBASSIGN

The Start Smart Guide *for Students*

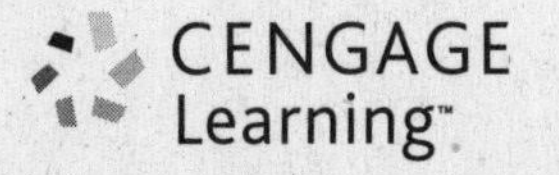

Australia • Brazil • Japan • Korea • Mexico • Singapore • Spain • United Kingdom • United States

CENGAGE Learning™

Enhanced WebAssign: The Start Smart Guide for Students

Acquisitions Editor: Gary Whalen

Copyeditor: Deborah Todd

Media Editor: Lynh Pham

Cover Design: Fabio Fernandes

WebAssign
Centennial Campus
730 Varsity Drive
Raleigh, NC 27606
Web: http://webassign.net
Tel: (800) 955-8275 or (919) 829-8181
Fax: (919) 829-1516
E-mail: info@webassign.net

For product information and technology assistance, contact us at **Cengage Learning Customer & Sales Support, 1-800-354-9706.**

For permission to use material from this text or product, submit all requests online at **www.cengage.com/permissions.**

Further permissions questions can be emailed to **permissionrequest@cengage.com.**

ISBN-13: 978-0-495-38479-3

ISBN-10: 0-495-38479-8

Cengage Learning is a leading provider of customized learning solutions with office locations around the globe, including Singapore, the United Kingdom, Australia, Mexico, Brazil, and Japan. Locate your local office at: **www.cengage.com/global**.

Cengage Learning products are represented in Canada by Nelson Education, Ltd.

To learn more about Cengage Learning, visit **www.cengage.com**.

Purchase any of our products at your local college store or at our preferred online store **www.cengagebrain.com.**

Printed in the United States of America
10 11 12 13 14 15 14 13 12 11

CONTENTS

WebAssign works with any recent browser and computer. Some assignments may require an updated browser and/or plugins like Java, Flash, Shockwave, or Adobe Reader.

For technical support go to http://webassign.net/student.html or email support@webassign.net.

Getting Started

Welcome to Enhanced WebAssign, the integrated, online learning system that gives you 24/7 access to your math, physics, astronomy, chemistry, biology, and statistics assignments.

Now, you can do homework, take quizzes and exams, and receive your scores and graded assignments from any computer with an Internet connection and web browser, any time of the day or night.

Note: As a live, web-based program, Enhanced WebAssign is updated regularly with new features and improvements. Please refer to WebAssign's online Help for the most current information.

Technical Startup Tips

Before you start, please note the following important points:

- ❍ Most standard web connections should work with WebAssign. We recommend using Firefox 1.0 or later, or Internet Explorer 5.5 or later. *We do not recommend the AOL browser.*
- ❍ You can use a 56 KBPS modem, broadband, or school network connection.
- ❍ Your browser needs to have both JavaScript and Java enabled.
- ❍ You *cannot skip the login page*. WebAssign must know it is you before delivering your assignments.

Note: If you'd like to bookmark WebAssign on your computer, we recommend that you bookmark **https://www.webassign.net/login.html** or the appropriate address for your school.

Login to WebAssign

In order to access WebAssign your instructor will provide you with login information or a Class Key. Login information will consist of a username, institution code, and an initial password. The Class Key will allow you to self-register and create your own login. You will need to remember the username and initial password you set after self-registering.

Please note that Class Keys are not the same as access codes. See pages 8–9 for instructions on registering your access code number. You will need to login first before you are able to register an access code.

➢ To get started

1. If you are using a shared computer, completely exit any browsers that are already open.
2. Open a new web browser and go to https://www.webassign.net/login.html, or the web address provided by your instructor.

   If your instructor has provided you with a **Username, Institution** (school code), and **Password,** continue with step 3. If you have been provided with a **Class Key** (usually your institution name and a series of numbers), skip to step 5.
3. Enter your **Username**, **Institution** (school code), and **Password** *provided by your instructor.*

   Institution

   If you do not know your **Institution,** you can search for it by clicking **(what's this?)** above the **Institution** entry box.

   In the **What's My Institution Code** pop-up window, enter your school name and click **go!**. The **Institution Search Results** table will give you choices of the School Names that most closely match your entry, and the **Institution Code** that you should enter in the **Institution** entry box on the **WebAssign Login** screen.

   Password

   If you have forgotten or do not know your **Password,** click **(Reset Password)** above the **Password** entry box, and follow the directions on the **WebAssign New Password Request** screen. You will need to submit your username, institution code, and the email address on file in your WebAssign account. If you are unsure of your username or listed email address, please check with your instructor. WebAssign cannot reset your username or password.

4. Click **Log In**.
5. If your instructor gave you a **Class Key,** you will use it to create your account. Click the **I have a Class Key** button. You will need to use this key only once when you register.
6. Enter the Class Key code in the field provided and click **Submit**. If your Class Key is recognized, you will be asked to confirm your class information. If the information provided is correct, then click the **Yes, this is my class** button. If the information provided is incorrect, then click the **No, this is not my class** button and reenter your Class Key.
7. When you verify that the class information is correct, you will be able to create a new WebAssign account. If you do not already have a WebAssign account, select "I need to create a WeAssign account." If you have a WebAssign account for any other course at the same institution then select "I already have a WebAssign account." When you have selected the appropriate option click the **Continue** button.
8. Enter a username in the field provided and then click **Check Availability** to determine whether or not your username is already in use. If it is, an available alternate username will be suggested. Remember your username because you will use it every time you login to WebAssign.
9. Enter and then re-enter a password. Remember your password because you will use it every time you login to WebAssign.
10. Under **Student Information** enter your first and last name, email address, and student ID.
11. Click **Create My Account**.
12. If you see confirmation of your account creation, you will now be able to login to WebAssign. Click **Log in now**.

Note: Before starting WebAssign on a shared computer, always exit any browsers and restart your browser application. *If you simply close the browser window or open a new window, login information contained in an encrypted key may not be yours.*

Logout

When you are finished with your work, click the **Logout** link in the upper right corner of your Home page, and *exit the browser completely* to avoid the possibility of someone else accessing your work.

YOUR ENHANCED WEBASSIGN HOME PAGE

Your personalized Home page is your hub for referencing and managing all of your Enhanced WebAssign assignments.

Using Access Codes

Some classes require an **access code** for admission. Please remember:

- An **access code** is *not* the same as a Class Key or a login password.
- An **access code** is good for *one class only* unless the textbook includes a two-term **access code**.
- An **access code** is an alphanumeric code that is *usually* packaged with your textbook. It can begin with 2 or 3 letters, followed by an alphanumeric code, or it can have a longer prefix such as **BCEnhanced-S** followed by four sets of four characters.
- If your textbook did not include an **access code**, you can buy one at your bookstore, or from your personalized Home page by clicking the **Purchase an access code online** button.

➢ To enter an access code

1. Under **WebAssign Notices**, select the proper prefix from the **Choose your access code prefix** pull-down menu.

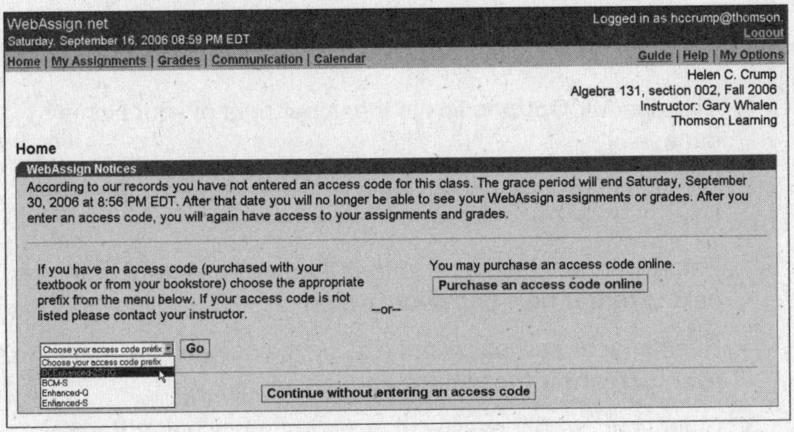

WebAssign.net
Saturday, September 16, 2006 08:59 PM EDT
Logged in as hccrump@thomson.
Logout
Home | My Assignments | Grades | Communication | Calendar
Guide | Help | My Options
Helen C. Crump
Algebra 131, section 002, Fall 2006
Instructor: Gary Whalen
Thomson Learning
Home
WebAssign Notices
According to our records you have not entered an access code for this class. The grace period will end Saturday, September 30, 2006 at 8:56 PM EDT. After that date you will no longer be able to see your WebAssign assignments or grades. After you enter an access code, you will again have access to your assignments and grades.
If you have an access code (purchased with your textbook or from your bookstore) choose the appropriate prefix from the menu below. If your access code is not listed please contact your instructor.
--or--
You may purchase an access code online.
Purchase an access code online
Choose your access code prefix
Go
Choose your access code prefix
BCEnhanced-2S/3Q
BCM-S
Enhanced-Q
Enhanced-S
Continue without entering an access code

WebAssign notices

2. Click **Go**.
3. In the entry boxes, type in your access code *exactly* as it appears on your card. (When you purchase online, the access code is entered automatically.)

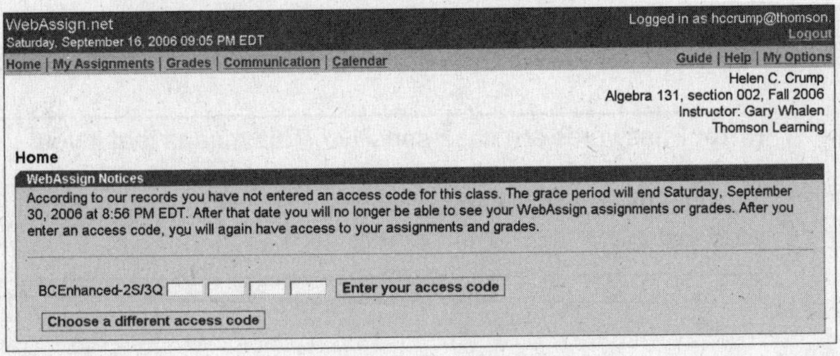

WebAssign.net
Saturday, September 16, 2006 09:05 PM EDT
Logged in as hccrump@thomson.
Logout
Home | My Assignments | Grades | Communication | Calendar
Guide | Help | My Options
Helen C. Crump
Algebra 131, section 002, Fall 2006
Instructor: Gary Whalen
Thomson Learning
Home
WebAssign Notices
According to our records you have not entered an access code for this class. The grace period will end Saturday, September 30, 2006 at 8:56 PM EDT. After that date you will no longer be able to see your WebAssign assignments or grades. After you enter an access code, you will again have access to your assignments and grades.
BCEnhanced-2S/3Q
Enter your access code
Choose a different access code

Access code entry

4. Click **Enter your access code.**

   If you have chosen the wrong prefix from the previous screen, you can click the **Choose a different access code** button to try again.

   If your **access code** is a valid unused code, you will receive a message that you have successfully entered the code for the class. Click the **Home** or **My Assignments** button to proceed.

Changing Your Password

For your personal security, it's a good idea to change the initial password provided by your instructor.

➢ To change your password

1. Click the **My Options** link in the upper right of your Home page.
2. In the **My Options** pop-up window, under the **Personal Info** tab:

   Enter your *new* password in the **Change Password** entry box next to **(enter new password)**, then

   Reenter your new password *exactly* the same in the entry box next to **(reenter for confirmation)**.
3. Enter your *current* password in the entry box under **If you made any changes above, enter your current password here and then click save:**, located at the bottom of the pop-up window.
4. Click the **Save** button in the bottom right corner of the pop-up window.

   If the change was successful, you will see the message **Your password has been changed**.

Note: Passwords are case-sensitive. This means that if you capitalize any of the letters, you must remember to capitalize them the same way each time you sign in to Enhanced WebAssign.

Changing Your Email Address

If your instructor provided you with an email address, you can easily change it to your own personal email address any time.

➢ To change your email address

1. Click the **My Options** link in the upper right of your Home page.

2. In the **My Options** pop-up window, under the **Personal Info** tab, enter your *valid* email address in the **Email Address** box.

3. Enter your current password in the entry box under **If you made any changes above enter your current password here and then click save:**, located at the bottom of the pop-up screen.

4. Click the **Save** button in the bottom right corner of the pop-up window.

   A confirmation email will be sent to your new email address.

Once you receive the confirmation email, you must click the link in the email to successfully complete and activate this change.

Working with Assignments

The courses that have been set up for you by your instructor(s) appear on your Enhanced WebAssign personalized Home page. If you have more than one course, simply select the course you want to work with from the pull-down menu.

Assignment Summary

There are two ways to get a quick summary of your assignments. On the Home page:

- Click the **My Assignments** link in the upper left *menu bar,* or
- Click the **Current Assignments** link in the **My Assignments** *module* on the Home page.

Accessing an Assignment

Once your assignments are displayed on your Home page, simply click the name of the assignment you'd like to begin.

- If you have previously submitted an assignment, you will see your most recent responses, if your instructor allows this feature.
- If you have already submitted the assignment, there will usually be a link to **Review All Submissions** on the page, if your instructor has allowed it.

WebAssign.net
Saturday, September 16, 2006 09:52 PM EDT
Logged in as hccrump@thomson.
Logout

Home | My Assignments | Grades | Communication | Calendar
Guide | Help | My Options

Helen C. Crump
Algebra 131, section 002, Fall 2006
Instructor: Gary Whalen
Thomson Learning

Home > My Assignments

Showing: Current Assignments | Past Assignments | All Assignments

Current Assignments

Assignment Name / Description	Score
In Class Presentations 1 (In Class) **Due: Monday, September 18, 2006 11:20 AM EDT** These presentations will be done in class on Monday and the scores uploaded by Wednesday.	**Score: 0**
HW3 Chapter 3 (Homework) **Due: Wednesday, September 20, 2006 09:00 PM EDT** Please read the notes from class and the first part of chapter 3. Then do this assignment. Please bring at least one practical application of what you learned to class.	**Score: 19 out of 21**
Quiz 2 (Quiz) 45 minutes allowed **Due: Friday, September 22, 2006 09:00 PM EDT** You will need to read sections 3.3 and 3.4 before starting this quiz. You will only have 45 minutes from the moment you open the quiz to complete the quiz and submit.	**Score: 0**

Assignment summary

Home > My Assignments > HW4 Chapter 4 (Homework)

Helen C. Crump
Algebra 131, section 002, Fall 2006
Instructor: Gary Whalen
Thomson Learning

About this Assignment

Due: **Wednesday, September 27, 2006 09:00 PM EDT**

Current Score: 0 out of 20
Question Score
Submission Options

Description
Please read sections 4.1 to 4.3 and make an outline. Then do the assignment. Afterwards find something interesting about the techniques you learned on Google.

1. --/6 points No Response | Show Details   Notes

Applications See the illustration below. A company manufactures various sizes of playpens having perimeters between 132 and 184 inches, inclusive.

(a) Complete the double inequality that describes the range of the perimeters of the playpens.

$\square \le 4s \le \square$

(b) Solve the double inequality to find the range of the side lengths of the playpens.

[?] [ ] , [ ] [?]

Read It

Submit New Answers To Question 1   Save Work

Math assignment

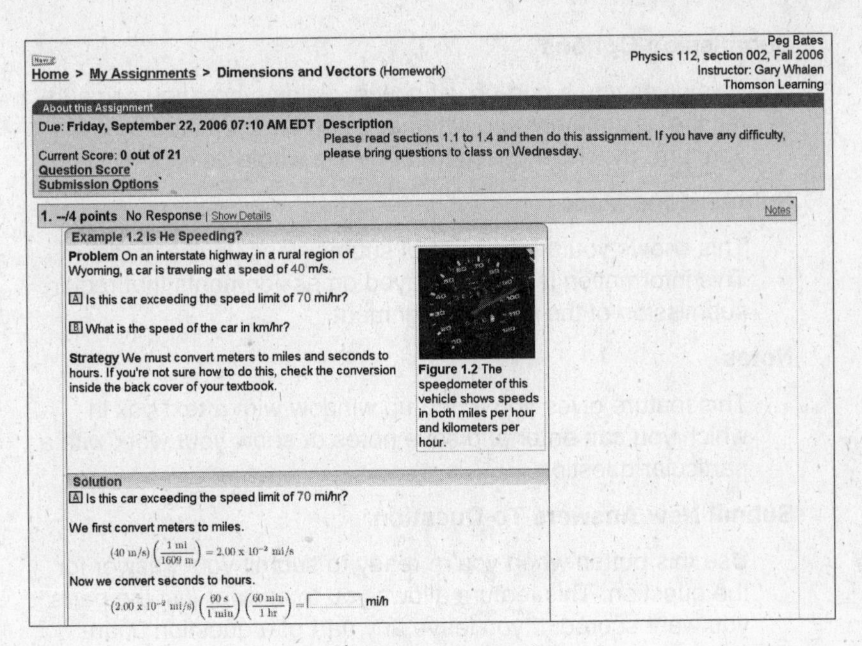

Physics assignment

Using the Assignment Page

When you click on an assignment name, your assignment will load. Within the **About this Assignment** page are links to valuable information about your assignment's score, submission options, and saving your work in progress. Within each question, there might also be "enhanced" action links to useful tutorial material such as book content, videos, animations, active figures, simulations, and practice problems. The links available may vary from one assignment to another.

Actions

Click a button or link to take one of the following actions:

Current Score

This gives you a quick look at your current score versus the maximum possible score.

Question Score

This gives you a pop-up window showing your score for each question.

Submission Options

This gives you a pop-up window explaining how you can submit the assignment and whether it can be submitted by question part, by whole question, or by the whole assignment.

Submissions Made

This shows you the number of submissions you've made. This information is only displayed on assignments that require submission of the entire assignment.

Notes

This feature gives you a pop-up window with a text box in which you can enter and save notes or show your work with a particular question.

Submit New Answers To Question

Use this button when you're ready to submit your answer for the question. This feature allows you to answer just the parts you want scored. If you leave any part of a question unanswered, the submission *will not* be recorded for that part.

Submit Whole Question

Use this button to submit your answer(s) for the entire question. If you leave any part of a question unanswered, the submission *will* be recorded as if the entire question has been answered, and will be graded as such.

Save Work

This button allows you to save the work you've done so far on a particular question, but does not submit that question for grading.

View Saved Work

Located in the question's header line, this link allows you to view work that you previously saved for that question.

Show Details

Located in the question's header line, this link shows your score on each part of the question, how many points each part of the question is worth, and how many submissions are allowed for each part if you can submit each part separately.

Submit All New Answers

This submits all of your new answers for all of the questions in the assignment.

Save All Work

This allows you to save all the work you've done on all of the questions in the assignment, but does not submit your work for grading.

Ask Your Teacher

This feature allows you to send a question about the assignment to your instructor.

Extension Request

This allows you to submit a request to your instructor for an extension of time on an assignment.

Home

This link takes you to your personalized Home page.

My Assignments

This link takes you to your assignments page.

Read it

This links to question-specific textbook material in PDF form.

Practice Another Version

This provides you with an alternate version of the assigned problem. Within the pop-up window you will be able to answer the practice problem and have that answer checked. You will also be able to practice additional versions of your assigned problem.

Practice it

This links to a practice problem or set of practice problems in a pop-up window. No grade is recorded on the work you do on practice problems.

Watch it

This links to a tutorial video.

Hint

This links to a pop-up window with helpful hints in case you get stuck on a question.

Hint: Active Figure

This links to an animated simulation to help you better understand the concepts being covered.

Note: Your instructor has the ability to turn on/off many of the options listed above.

ANSWERING QUESTIONS

Enhanced WebAssign uses a variety of question types that you're probably already familiar with using, such as multiple choice, true/false, free response, etc.

Always be sure to pay close attention to any instructions within the question regarding how you are supposed to submit your answers.

Numerical Questions

There are a few key points to keep in mind when working on numerical questions:

- Numbers can be entered in both scientific notation and numerical expressions, such as fractions.
- WebAssign uses the standard scientific notation "E" or "e" for "times 10 raised to the power." (Note: both uppercase E and lowercase e are acceptable in WebAssign.) For example, 1e3 is the scientific notation for 1000.
- Numerical answers may not contain commas (,) or equal signs (=).
- Numerical answers may only contain:
    - Numbers
    - E or e for scientific notation
    - Mathematical operators +, -, *, /
- Numerical answers within 1% of the actual answer are counted as correct, unless your instructor chooses a different tolerance. This is to account for rounding errors in calculations. In general, enter three significant figures for numerical answers.

➢ Example: Numerical Question

Let's suppose you're presented a question to which your answer is the fraction "one over sixty-four." Following are examples of Correct and Incorrect answer formats:

Correct Answers

Any of these formats would be correct:

1/64

0.015625

0.0156

.0156

1.5625E-2

Incorrect Answers

These formats would be graded as incorrect:

O.015625	The first character is the letter "O"
0. 015625	There is an improper space in the answer
1.5625 E-2	There is an improper space using E notation
l/64	The first character is lowercase letter "L"
5,400	There is a comma in the answer
1234.5=1230	There is an equal sign in the answer

Numerical Questions with Units

Some Enhanced WebAssign questions require a number and a unit, and this is generally, although not always, indicated in the instructions in the question.

You will know that a unit is expected when there is no unit after the answer box.

When you are expected to enter units and do not, you will get an error message telling you that units are required.

Note: Whether omission of the unit counts as a submission depends on the submission options chosen by the instructor.

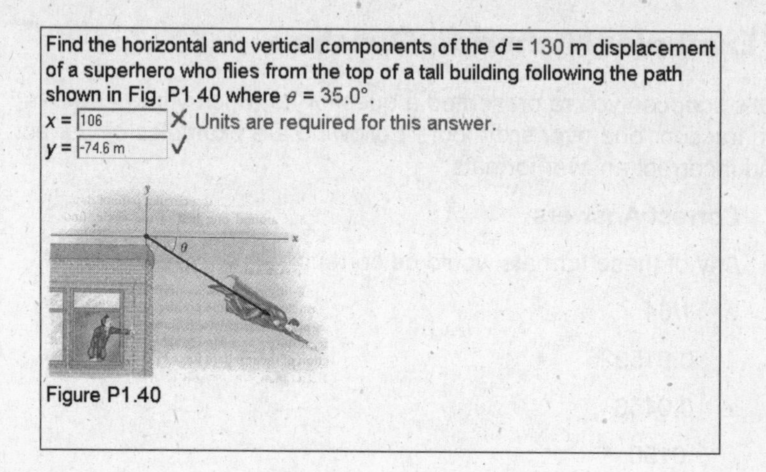

Numerical with units

The easiest units to use in this question are m, but the answer converted to yd would also be scored correct.

Numerical Questions with Significant Figures

Some numerical questions require a specific number of significant figures (sig figs) in your answer. If a question checks sig figs, you will see a sig fig icon next to the answer box.

If you enter the correct value with the wrong number of sig figs, you will not receive credit, but you will receive a hint that your number does not have the correct number of sig figs. The sig fig icon 4.0✓ is also a link to the rules used for sig figs in WebAssign.

Carry out the following arithmetic operations. (Use the correct number of significant figures.)

(a) the sum of the measured values 760., 37.2, 0.81, and 2.2
4.0✓ 8e2 ✗ Check the number of significant figures.

(b) the product 3.4 × 3.563
4.0✓ 12 ✓

(c) the product 5.7 × π
4.0✓ 18 ✓

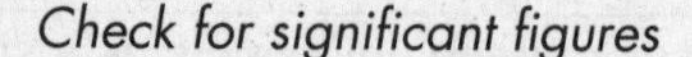
Check for significant figures

Math Notation: Using the MathPad

In many math questions, Enhanced WebAssign gives you a **MathPad**. The **MathPad** provides easy input of math notation and symbols, even the more complicated ones. If your answer involves math notation or symbols, the MathPad will become available when you click the answer box.

1

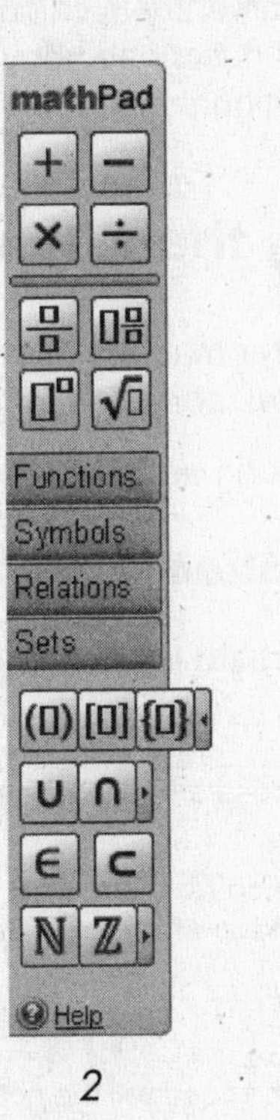

2

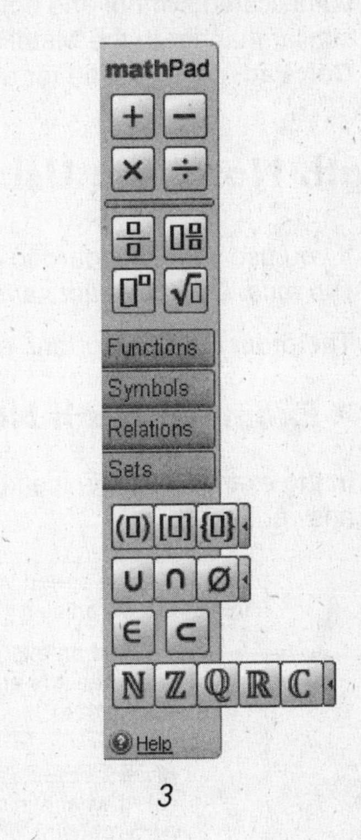

3

Top Symbols

The buttons on the top are single input buttons for frequently used operations.

Word Buttons

When you click one of the word buttons **Functions**, **Symbols**, **Relations**, or **Sets**, you are given a drop-down menu with symbols or notation from which to choose. For example, if you click the **Sets** button, you get set notation (figure 2 above). If you then click a right arrow button, additional symbols become available (figure 3 above).

To insert any available notation or symbol into your answer, simply click it.

Math Notation: Using the CalcPad

CalcPad, as its name implies, is designed for use with the more complicated symbol and notation entry in calculus. It functions in a similar manner to the MathPad described above. If your course uses **CalcPad**, check online for additional instructions.

Math Notation: Using the Keyboard

If you use your keyboard to enter math notation (calculator notation), *you must use the exact variables specified in the questions.*

The order is not important, as long as it is mathematically correct.

➢ Example: Math Notation Using Keyboard

In the example below, the keyboard is used to enter the answer in the answer field.

A car moves at speed *v* across a bridge made in the shape of a circular arc of radius *r*.

(a) Find an expression for the normal force acting on the car when it is at the top of the arc. (Use *m*, *g*, *v*, and *r* as appropriate.)

mg-mv^2/r

symbolic formatting help

(b) At what minimum speed will the normal force become zero (causing occupants of the car to seem weightless) if *r* = 23.5 m?

m/s

Symbolic question

Expression Preview

Clicking the eye button allows you to preview the expression you've entered in calculator notation.

Use this preview feature to help determine if you have properly placed your parentheses.

Symbolic Formatting Help

If you're unsure about how to symbolically enter your answer properly, use the **symbolic formatting help** button to display allowed notation.

Allowed Notation for Symbolic Formatting

+ for addition	x+1
- for subtraction	x-1, or -x
* or nothing for multiplication	4*x, or 4x
/ for division	x/4
** or ^ for exponential	x**3, or x^3
( ) where necessary to group terms	4/(x+1), or 3(x+1)
abs( ) to take the absolute value of a variable or expression	abs(-5) = 5
sin, cos, tan, sec, csc, cot, asin, acos, atan functions (angle x expressed in radians)	sin(2x)
sqrt( ) for square root of an expression	sqrt(x/5)
x^ (1/n) for the nth root of a number	x^ (1/3), or (x-3)^ (1/5)
pi for 3.14159...	2 pi x
e for scientific notation	1e3 = 1000
ln( ) for natural log	ln(x)
exp( ) for "e to the power of"	exp(x) = e^x

USING THE GRAPHING UTILITY

The Enhanced WebAssign Graphing Utility lets you graph one or more mathematical elements directly on a set of coordinate axes. Your graph is then scored automatically when you submit the assignment for grading.

The Graphing Utility currently supports points, rays, segments, lines, circles, and parabolas. Inequalities can also be indicated by filling one or more areas.

Graphing Utility Interface Overview

The middle of the Graphing Utility is the drawing area. It contains labeled coordinate axes, which may have different axis scales and extents, depending on the nature of the question you are working on.

On the left side of Graphing Utility is the list of Tools that let you create graph objects and select objects to edit.

The bottom of the Graphing Utility is the Object Properties Toolbar, which becomes active when you have a graph element selected. This toolbar shows you all the details about the selected graph object, and also lets you edit the object's properties.

On the right side of Graphing Utility is a list of Actions that lets you create fills and delete objects from your graph.

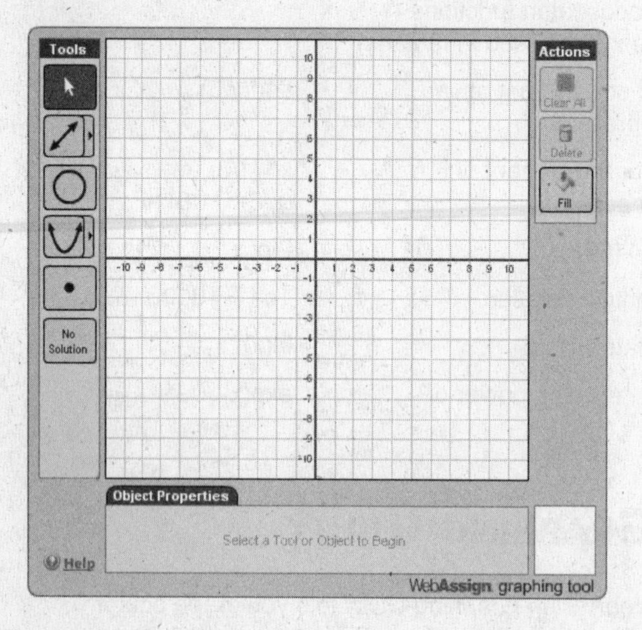

Drawing Tools	
Points:	Click the Point Tool, and then click where you want the point to appear.
Lines:	Click the Line Tool, and then place two points along the desired line. The arrows on the end of the line indicate that the line goes off to infinity on both ends.
Rays:	Click the Ray Tool, place the endpoint, and then place another point on the ray.
Line Segments:	Click the Line Segment Tool, and then place the two endpoints of the line segment.
Circles:	Click the Circle Tool, place the point at the center first, and then place a point on the circumference of the circle.
Parabolas:	Click the Parabola Tool (either vertical or horizontal), place the vertex first, and then place another point on the parabola.
No Solution: No Solution	If the question has no solution, simply click the No Solution Tool.

Note: Don't worry if you don't place the points exactly where you want them initially; you can move these points around before submitting for grading.

Selecting Graph Objects

To edit a graph object, it must be "selected" as the active object. (When you first draw an object, it is created in the selected state.) When a graph element is "selected," the color of the line changes and two “handles” are visible. The handles are the two square points you clicked to create the object. To select an object, click the Select Tool, and then click on the object. To deselect the object, click on a blank area on the drawing area or on a drawing tool.

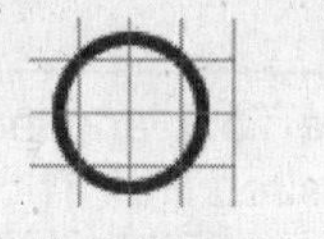

Not Selected

Selected

Once an object is selected, you can modify it by using your mouse or the keyboard. As you move it, you’ll notice that you cannot move the handles off the drawing area. To move an object with the mouse, click and drag the object’s line. Or, click and drag one of the handles to move just that handle. On the keyboard, the arrow keys can also move the selected object by one unit at a time.

As you move the object or handle you’ll see that the Object Properties toolbar reflects your changes.

Object Properties Toolbar	
Coordinate Fields: Point 1 (-13 , -9) Point 2 (15 , -6)	You can use the Coordinate Fields to edit the coordinates of the handles directly. Use this method to enter decimal or fractional coordinates.
Endpoint Controls: Endpoint Endpoint	If the selected object is a segment or ray, the Endpoint Controls are used to define the endpoint as closed or open. As a shortcut, you can also define an endpoint by clicking on the endpoint when the ray or segment is in the unselected state.
Solid/Dash Controls: Solid Dash	For any selected object other than a point, the Solid/Dash Controls are used to define the object as solid or dashed. To change graph objects to solid or dashed, select the object and then click the **Solid** or **Dash** button.

Using Fractions or Decimals as Coordinates

To draw an object with handle coordinates that are fractions or decimals, you must use the Object Properties Toolbar. Draw the desired object anywhere on the drawing area, then use the Coordinate Fields to change the endpoint(s) to the desired value. For example, to enter a fraction, just type "3/4."

Note: The points and lines you draw must be exactly correct when you submit for grading. This means you should not round any of your values—if you want a point at 11/3, you should enter 11/3 in the coordinate box rather than 3.667. Also, mixed fractions are not acceptable entries. This means that 3 2/3 is an incorrect entry.

Actions	
Fill Tool: Fill	To graph an inequality, you must specify a region on the graph. To do this, first draw the line(s), circle(s), or other object(s) that will define the region you want to represent your answer. Be sure to specify the objects as either solid or dashed, according to the inequality you are graphing! Then choose the Fill Tool and click inside the region that you want filled. -4 -5 -6 -7 -8 -9 -10 If you decide you want the fill in a different area, click the filled region that you want to unfill, and then click the region that you do want to fill.
Delete Tool: Delete	To erase a single graph object, first select that element in the drawing area, then click the Delete Tool or press the Delete key on your keyboard.

Actions	
Clear All Tool: Clear All	The Clear All Tool will erase all of your graph objects. (If the drawing area is already empty, the Clear All Tool is disabled.)

➢ Example

Let's suppose you're asked to graph the inequality $y > 5x + \frac{1}{5}$, and you want to use the points $\left(0, \frac{1}{5}\right)$ and $\left(1, 5\frac{1}{5}\right)$. First, you would place any line on the drawing area.

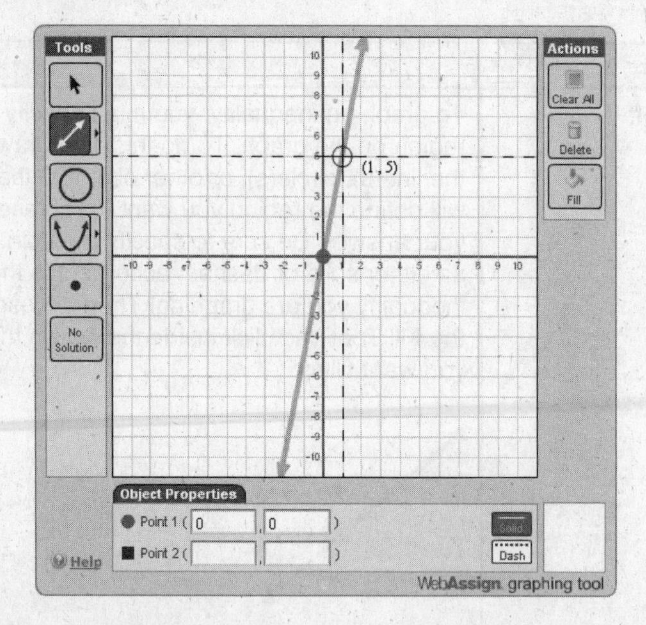

Then, you would adjust the points using the Coordinate Fields. Remember, you need to enter 5 1/5 as a decimal (5.2) or an improper fraction (26/5).

Object Properties
Point 1 (0 , 1/5)
Point 2 (1 , 5.2)
Solid
Dash
Help
WebAssign graphing tool

Next, you would define the line as dashed since the inequality does not include the values on the line.

Finally, you would select the Fill Tool and click on the desired region to complete the graph.

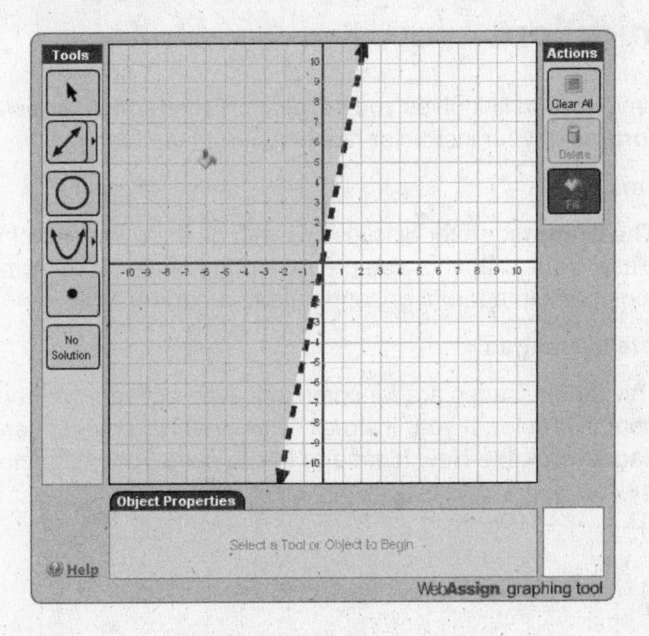

Additional Features

Calendar

The Calendar link presents you with a calendar showing all of your assignments on their due dates. You can also click on any date and enter your own personal events.

Communication

The Communication link gives you access to **Private Messages** and course **Forums**, if your instructor has enabled these features.

Forums

The **Forums** are for discussions with all the members of your class. Your instructor can create forums, and you can create topics within a forum or contribute to a current topic.

Private Messages

Private Messages are for communication between you and your instructor. If your instructor has enabled private messages, click the **New Message** link to send your instructor a message.

Grades

The **Grades** link at the top of all your WebAssign pages gives you access to the raw scores and grades that your instructor posts. This page may also include statistics on the whole class, and a histogram of scores for each category of assignment and each individual assignment. It may have your individual average for each category of assignment, as well as the score on each of your assignments.

Your instructor will let you know what Scores and Grades will be posted in your course.

If your instructor has enabled all of the options, your display will be similar to the one below.

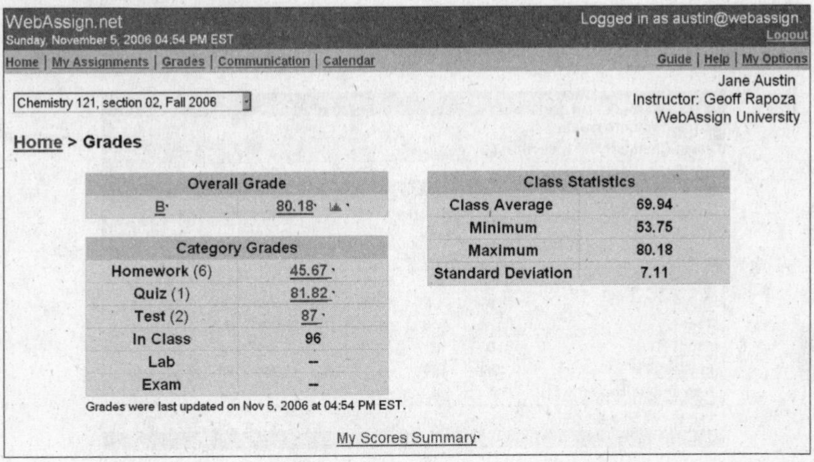

WebAssign.net
Sunday, November 5, 2006 04:54 PM EST
Logged in as austin@webassign.
Logout

Home | My Assignments | Grades | Communication | Calendar
Guide | Help | My Options

Chemistry 121, section 02, Fall 2006

Jane Austin
Instructor: Geoff Rapoza
WebAssign University

Home > Grades

Overall Grade	
B	80.18

Category Grades	
Homework (6)	45.67
Quiz (1)	81.82
Test (2)	87
In Class	96
Lab	–
Exam	–

Grades were last updated on Nov 5, 2006 at 04:54 PM EST.

Class Statistics	
Class Average	69.94
Minimum	53.75
Maximum	80.18
Standard Deviation	7.11

My Scores Summary

Grades

Overall Grade

This score is calculated from the various categories of assignments—for example, **Homework, Test**, **In Class**, **Quiz**, **Lab**, and **Exam**. Your instructor may have different categories.

Category Grades

The **Category Grades** give the contribution to your overall grade from each of the categories. If you click a grade that is a link, you will get a pop-up window explaining how the number was calculated.

Class Statistics

Class Statistics shows the averages, minimum scores, maximum scores, and standard deviation of the class at large.

My Scores Summary

This link presents a pop-up window with a summary of your raw scores and the class statistics on each assignment, if your teacher has posted these.

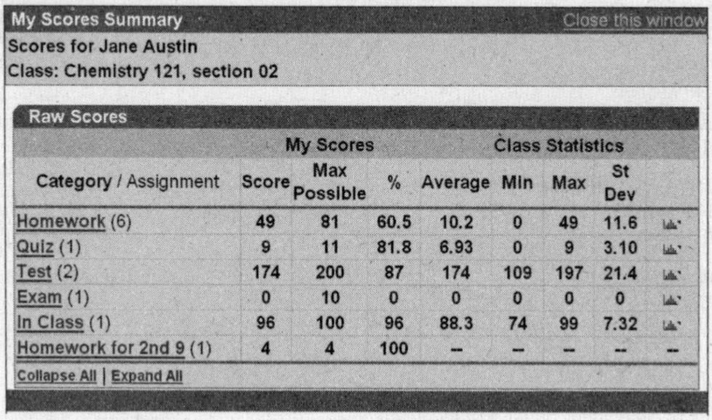

My Scores Summary Close this window
Scores for Jane Austin
Class: Chemistry 121, section 02

Raw Scores

	My Scores			Class Statistics				
Category / Assignment	Score	Max Possible	%	Average	Min	Max	St Dev	
Homework (6)	49	81	60.5	10.2	0	49	11.6	
Quiz (1)	9	11	81.8	6.93	0	9	3.10	
Test (2)	174	200	87	174	109	197	21.4	
Exam (1)	0	10	0	0	0	0	0	
In Class (1)	96	100	96	88.3	74	99	7.32	
Homework for 2nd 9 (1)	4	4	100	--	--	--	--	--

Collapse All | Expand All

My Scores summary

Technical Tips

Enhanced WebAssign relies on web browsers and other related technology that can lead to occasional technical issues. The following technical tips can help you avoid some common problems. .

Cookies

Allow your browser to accept cookies.

WebAssign will work if you set your browser to not accept cookies; however, if an encrypted cookie is not saved to your computer during your session, you may be asked to login again more frequently. Once you logout and exit your browser, the cookie is deleted.

For technical support go to http://webassign.net/student.html or email support@webassign.net.

Login and Credit

If you see an assignment that does not have your name at the top, you have not logged in properly.

You will not receive credit for any work you do on an assignment if your name is not associated with it. If you find yourself in the midst of this situation, make notes of your solution(s) and start over. Be aware that any randomized values in the questions will probably change.

Logout When You Finish Your Session

If someone wants to use your computer for WebAssign, logout and exit the browser before relinquishing control.

Otherwise, the work you have just completed may be written over by the next user.

Server

Although it is very rare, the WebAssign server may occasionally be unavailable.

If the WebAssign server is unavailable, instructors will provide instructions for submitting your assignments—possibly including new due dates. The policy for handling server problems will vary from instructor to instructor.

Use the Latest Browser Software

Use the latest version of Firefox, Mozilla, Netscape, or Internet Explorer browsers.

Older versions of browsers may not be supported by WebAssign.

For technical support go to http://webassign.net/student.html or email support@webassign.net.